EV CHARGERS LEVEL 2 INSTALLATION GUIDE

Dr. Maxwell Shimba

Shimba Publishing, LLC.

Printed by Shimba Publishing LLC
Printed in the United States of America

TABLE OF CONTENTS

INTRODUCTION

The adoption of electric vehicles (EVs) is accelerating at an unprecedented rate, driven by the increasing awareness of environmental issues, advancements in battery technology, and the push for cleaner transportation alternatives. As more people make the switch from traditional internal combustion engine vehicles to electric ones, the infrastructure to support these vehicles, particularly charging stations, becomes critically important. Among the various types of EV chargers available, Level 2 chargers stand out for their balance of speed and practicality, making them a popular choice for both residential and commercial installations.

Why Level 2 Charging?

Level 1 chargers, which plug into standard 120-volt outlets, offer a convenient but slow charging option. They typically provide around 4-5 miles of range per hour of charging, which can be sufficient for plug-in hybrid vehicles or for overnight charging if daily driving distances are minimal. However, for most EV owners, especially those with longer commutes or limited charging windows, Level 1 chargers may not be adequate.

Level 2 chargers, on the other hand, operate at 240 volts and can deliver between 10 to 60 miles of range per hour, depending on the vehicle and charger specifications. This significant increase in charging speed makes Level 2 chargers an ideal solution for daily home use, workplace charging, and public charging stations. They strike a perfect balance between cost, installation complexity, and charging efficiency.

Understanding the Benefits of Level 2 Chargers

1. Faster Charging Times: Level 2 chargers can recharge an EV's battery much more quickly than Level 1 chargers. This is especially beneficial for those who drive long distances daily or have multiple EVs at home.

2. Increased Convenience: With faster charging, EV owners can more easily fit charging into their daily routines without worrying about long downtimes. This makes electric vehicle ownership more practical and appealing.

3. Enhanced Battery Health: Some studies suggest that slower, consistent charging (which Level 2 chargers can provide compared to the very high speeds of DC fast chargers) may help prolong the lifespan of an EV's battery.

4. Future-Proofing: As battery technology continues to improve, newer EV models are expected to have larger batteries and higher range. Investing in a Level 2 charger now ensures that your charging infrastructure will remain relevant and capable of handling these advancements.

The Scope of This Guide

This guide aims to provide a comprehensive overview of the Level 2 EV charger installation process. Whether you are a homeowner looking to install a charger in your garage, a business owner planning to provide EV charging for employees and customers, or an installer seeking a detailed reference, this guide covers all the necessary steps and considerations.

We will walk you through:

- Understanding Level 2 Chargers: Detailed information about what Level 2 Chargers are, their benefits, and the different types available.

- Assessing Your Charging Needs: How to determine your specific charging requirements based on your driving habits, vehicle type, and other factors.

- Electrical Requirements and Safety: Key electrical specifications, safety standards, and codes that must be adhered to for a safe and compliant installation.

- Choosing the Right Level 2 Charger: Features to consider when selecting a charger and reviews of popular brands and models.

- Installation Process: A step-by-step guide to installing your Level 2 charger, including preparations, tools, materials, and detailed instructions for both professional and DIY installations.

- Post-Installation Considerations: Tips for testing, troubleshooting, regular maintenance, and potential upgrades.

- Incentives and Rebates: Information on federal, state, and local incentives, and how to apply for them.

- Future Trends in EV Charging: A look at emerging technologies and their potential impact on EV charging infrastructure.

Embracing the EV Revolution

The shift to electric vehicles is more than just a change in technology; it's a movement towards a more sustainable and eco-friendly future. By installing a Level 2 charger, you are not only enhancing your own convenience and vehicle performance but also contributing to the broader effort of reducing carbon emissions and dependency on fossil fuels.

As you embark on this journey, this guide will serve as a valuable resource, providing the knowledge and tools needed to make informed decisions and ensure a successful installation. Welcome to the future of transportation – clean, efficient, and electric.

xi

DR. MAXWELL SHIMBA

CHAPTER 01

1.1 UNDERSTANDING LEVEL 2 CHARGERS

1.1 What is a Level 2 Charger?

Level 2 chargers represent a significant upgrade from the basic Level 1 charger in terms of power and efficiency. Operating at 240 volts, Level 2 chargers are designed to charge electric vehicles (EVs) much faster than their 120-volt Level 1 counterparts. This chapter delves into the specifics of what makes Level 2 chargers unique, their benefits, and their various applications in different settings.

1.2 How Level 2 Chargers Work

To understand how Level 2 chargers work, it's essential to first grasp the basics of EV charging. Electric vehicles use onboard chargers to convert the AC power from the charger into DC power that is stored in the vehicle's battery. Level 2 chargers supply this AC power at 240 volts, allowing for a higher current flow and, consequently, faster charging times.

Technical Specifications

- Voltage: 240V (typically found in residential and commercial electrical systems)

- Amperage: Ranges from 16 to 80 amps, with most residential units operating between 30 to 40 amps

- Power Output: Typically between 3.3 kW and 19.2 kW, depending on the charger and vehicle capabilities

1.3 Benefits of Level 2 Charging

Faster Charging Times

One of the most significant advantages of Level 2 chargers is their ability to recharge an EV much more quickly than Level 1 chargers. For example, a typical Level 1 charger provides about 4-5 miles of range per hour of charging. In contrast, a Level 2 charger can deliver between 10 to 60 miles of range per hour, depending on the vehicle and charger specifications. This means that an EV with a typical 60 kWh battery can be fully charged in 4 to 10 hours using a Level 2 charger, compared to 20 to 30 hours with a Level 1 charger.

Increased Convenience

The faster charging times offered by Level 2 chargers translate into greater convenience for EV owners. Instead of needing to leave the vehicle plugged in for an entire day, users can achieve a full charge overnight or even during a typical workday. This makes it much easier to integrate EV charging into daily routines, reducing downtime and enhancing the overall user experience.

Enhanced Battery Health

While DC fast chargers provide the quickest charging times, they can generate more heat and put additional stress on the battery, potentially reducing its lifespan over time. Level 2 chargers offer a balanced approach, providing a fast but gentle charge that is less likely to degrade the battery. Regularly using a Level 2 charger can help maintain the health and longevity of the EV's battery, ensuring optimal performance over the long term.

1.4 Types of Level 2 Chargers

Level 2 chargers come in various forms to suit different installation environments and user needs. Here are the primary types:

Wall-Mounted Units

These are the most common type of Level 2 chargers for residential use. They are mounted on the wall of a garage or an exterior wall of a home, providing a convenient and space-saving solution. Wall-mounted units typically include a charging cable that can be easily wrapped and stored when not in use.

Pedestal-Mounted Units

Pedestal-mounted units are often found in commercial or public charging stations. These units are free-standing and can be installed in parking lots, garages, or other areas where wall space may not be available. Pedestal units are robust and designed to withstand the elements, making them suitable for outdoor installations.

Portable Level 2 Chargers

Portable Level 2 chargers offer flexibility and convenience for EV owners who need to charge their vehicles in multiple locations. These units are compact and can be easily transported, allowing users to take them on trips or use them in different locations. They typically come with adapters to connect to various types of 240-volt outlets, such as those used for RVs or heavy-duty appliances.

1.5 Installation Considerations

Electrical Panel Capacity

Before installing a Level 2 charger, it's crucial to ensure that your electrical panel can support the additional load. Most Level 2 chargers require a dedicated 240-volt circuit, which may necessitate upgrading your electrical panel or adding a subpanel. Consulting with a licensed electrician can help determine if your current electrical system is adequate and what modifications, if any, are needed.

Proximity to Parking Space

When selecting a location for your Level 2 charger, consider the proximity to where you park your EV. The charger should be easily accessible and within reach of the vehicle's charging port. For wall-mounted units, ensure that there is sufficient space on the wall and that the charging cable can comfortably reach the vehicle. For pedestal-mounted units, choose a location that allows for convenient access without obstructing walkways or other vehicles.

Environmental Considerations

If you plan to install a Level 2 charger outdoors, it's essential to choose a unit that is designed for outdoor use and can withstand various weather conditions. Look for chargers with durable, weather-resistant casings and ensure that the installation site provides adequate protection from rain, snow, and extreme temperatures. Additionally, consider the length of the charging cable and whether it will remain flexible and easy to use in cold weather.

Understanding the capabilities and benefits of Level 2 chargers is the first step in making an informed decision about your EV charging needs. Level 2 chargers offer a practical and efficient solution for both residential and commercial installations, providing faster charging times, increased convenience, and enhanced battery health. By selecting the right type of charger and ensuring proper installation, you can enjoy the many advantages of Level 2 charging and make the most of your electric vehicle experience.

In the next chapter, we will delve into assessing your specific charging needs, helping you determine the best charger and installation setup based on your driving habits, vehicle type, and other essential factors.

1.2 BENEFITS OF LEVEL 2 CHARGING

Electric vehicles (EVs) have revolutionized the automotive industry, offering a sustainable alternative to traditional internal combustion engine vehicles. As the number of EVs on the road continues to grow, so does the demand for efficient and convenient charging solutions. Level 2 chargers, which operate at 240 volts, have emerged as a popular choice due to their ability to charge EVs significantly faster than standard Level 1 chargers. This chapter explores the various benefits of Level 2 charging, including faster charging times, increased convenience for daily use, and enhanced battery life and performance.

Faster Charging Times

One of the most compelling advantages of Level 2 chargers is their ability to charge EVs much more quickly than Level 1 chargers. While a standard Level 1 charger, operating at 120 volts, provides about 4-5 miles of range per hour of charging, a Level 2 charger can deliver between 10 to 60 miles of range per hour, depending on the vehicle and charger specifications.

Comparison of Charging Speeds

- Level 1 Charger: Adds approximately 4-5 miles of range per hour.

- Level 2 Charger: Adds approximately 10-60 miles of range per hour.

This significant increase in charging speed means that an EV with a typical 60 kWh battery can be fully charged in 4 to 10 hours using a Level 2 charger, compared to 20 to 30 hours with a Level 1 charger. The faster charging times offered by Level 2 chargers make them an ideal solution for both residential and commercial settings.

Increased Convenience for Daily Use

The faster charging times provided by Level 2 chargers translate into greater convenience for EV owners. Instead of needing to leave the vehicle plugged in for an entire day, users can achieve a full charge overnight or even during a typical workday. This makes it much easier to integrate EV charging into daily routines, reducing downtime and enhancing the overall user experience.

Benefits of Faster Charging

- Overnight Charging: EV owners can charge their vehicles overnight and wake up to a fully charged battery, ready for the day ahead.

- Workplace Charging: Employees can charge their vehicles during the workday, ensuring they have sufficient range for their commute home and any additional driving needs.

- Public Charging Stations: Faster charging times at public stations mean that EV owners spend less time waiting for their vehicles to charge, making long trips more feasible and convenient.

The increased convenience of Level 2 charging also supports the growing trend of multi-vehicle households. With multiple EVs to charge, the faster charging speeds of Level 2 chargers ensure that all vehicles can be adequately charged without causing conflicts or scheduling issues.

Enhanced Battery Life and Performance

While DC fast chargers provide the quickest charging times, they can generate more heat and put additional stress on the battery, potentially reducing its lifespan over time. Level 2 chargers offer a balanced approach, providing a fast but gentle charge that is less likely to degrade the battery.

Regularly using a Level 2 charger can help maintain the health and longevity of the EV's battery, ensuring optimal performance over the long term.

Battery Health Considerations

- Reduced Heat Generation: Level 2 chargers generate less heat compared to DC fast chargers, which helps protect the battery from thermal degradation.

- Consistent Charging: Level 2 chargers provide a steady and consistent charge, which can be better for battery health compared to the high power levels of DC fast chargers.

- Optimal Charging Speed: While Level 1 chargers are slower and gentler, they may not always provide sufficient charging speed for daily needs. Level 2 chargers strike a balance, offering a speed that is fast enough for daily use but gentle enough to avoid significant wear on the battery.

By using a Level 2 charger for regular charging, EV owners can ensure that their batteries remain in good condition, leading to better overall performance and a longer lifespan for their vehicles. This not only enhances the driving experience but also helps protect the investment made in the vehicle.

The benefits of Level 2 charging are clear: faster charging times, increased convenience for daily use, and enhanced battery life and performance. These advantages make Level 2 chargers a practical and efficient solution for both residential and commercial installations. By understanding these benefits, EV owners can make informed decisions about their charging needs and enjoy a more seamless and enjoyable electric vehicle experience.

In the next chapter, we will delve into assessing your specific charging needs, helping you determine the best charger and installation setup based on your driving habits, vehicle type, and other essential factors.

1.3 TYPES OF LEVEL 2 CHARGERS

Level 2 chargers are available in various forms to cater to different needs and installation environments. Choosing the right type of Level 2 charger is essential for maximizing convenience, functionality, and safety. In this chapter, we will explore the three primary types of Level 2 chargers: wall-mounted units, pedestal-mounted units, and portable Level 2 chargers. Each type has its own set of features and advantages, making it suitable for specific applications and user requirements.

Wall-Mounted Units

Wall-mounted units are the most common type of Level 2 chargers, especially for residential use. These chargers are designed to be installed on the wall of a garage, carport, or exterior wall of a home. They offer a convenient and space-saving solution for EV owners who want to charge their vehicles at home.

Features and Advantages

- Space-Saving Design: Wall-mounted units take up minimal space and can be installed in areas where floor space is limited. This makes them ideal for home garages or carports.

- Ease of Use: These units typically come with a charging cable that can be easily wrapped and stored when not in use. Many models also feature a cable management system to keep the charging area neat and organized.

- Cost-Effective: Wall-mounted units are generally more affordable than pedestal-mounted units, making them a cost-effective choice for residential installations.

- Customizable Installation: Homeowners can choose the most convenient location for installation, ensuring that the charger is easily accessible and that the cable can comfortably reach the vehicle's charging port.

Considerations

- Mounting Surface: Ensure that the wall where the unit will be mounted is sturdy and capable of supporting the charger's weight. Additional support may be needed for some installations.

- Weather Protection: If the unit is installed outdoors, it should be weather-resistant to withstand various environmental conditions. Look for chargers with durable casings and proper sealing to protect against moisture and dust.

Pedestal-Mounted Units

Pedestal-mounted units are commonly used in commercial or public charging stations. These chargers are free-standing and can be installed in parking lots, garages, or other areas where wall space may not be available. Pedestal-mounted units are designed to be robust and durable, making them suitable for high-traffic and outdoor environments.

Features and Advantages

- Durability: Pedestal-mounted units are built to withstand the rigors of public use and adverse weather conditions. They are typically made from heavy-duty materials and have a rugged design.

- Accessibility: These units are often installed in locations that provide easy access for multiple users, such as public parking lots, workplaces, and multi-family residential complexes. The free-standing design allows for flexible placement and easy access from multiple directions.

- Scalability: Pedestal-mounted units are ideal for installations where multiple chargers are needed. They can be arranged in rows or clusters to accommodate a large number of EVs simultaneously.

- Professional Appearance: The sleek and professional appearance of pedestal-mounted units makes them a good fit for commercial and public spaces, enhancing the overall look of the charging area.

- Installation Cost: Pedestal-mounted units can be more expensive to purchase and install compared to wall-mounted units. The installation may require additional groundwork, such as pouring concrete bases and running underground electrical conduit.

- Space Requirements: Ensure that there is sufficient space for the pedestal and that the charging cables can reach the vehicles without creating tripping hazards or obstructing walkways.

Portable Level 2 Chargers

Portable Level 2 chargers offer flexibility and convenience for EV owners who need to charge their vehicles in multiple locations. These units are compact and can be easily transported, allowing users to take them on trips or use them in different locations.

Features and Advantages

- Flexibility: Portable Level 2 chargers can be used at home, at work, or on the go. They are perfect for EV owners who travel frequently or who do not have a dedicated charging station at home.

- Compatibility: Many portable chargers come with adapters to connect to various types of 240-volt outlets, such as those used for RVs or heavy-duty appliances. This makes them versatile and compatible with a wide range of power sources.

- Ease of Storage: When not in use, portable chargers can be easily stored in the trunk of the vehicle or in a storage compartment at home. Their compact size makes them convenient to carry and store.

Considerations

- Charging Speed: While portable Level 2 chargers offer faster charging than Level 1 chargers, they may not be as fast as some wall-mounted or pedestal-mounted units. Check the specifications to ensure that the charging speed meets your needs.

- Durability: Because they are designed to be portable, these chargers may not be as rugged as permanent installations. Ensure that the unit is sturdy and has protective features to withstand frequent handling and transportation.

- Availability of Power Outlets: Portable chargers require access to a suitable 240-volt power outlet. When planning to use the charger on the go, make sure that compatible outlets are available at your destination.

Choosing the right type of Level 2 charger depends on your specific needs and installation environment. Wall-mounted units are ideal for residential settings, offering a cost-effective and space-saving solution. Pedestal-mounted units are perfect for commercial and public spaces, providing durability and scalability. Portable Level 2 chargers offer unmatched flexibility, making them a great choice for EV owners who need to charge their vehicles in multiple locations.

By understanding the features and advantages of each type of Level 2 charger, you can make an informed decision and select the charger that best meets your needs. In the next chapter, we will delve into assessing your specific charging requirements, helping you determine the best charger and installation setup based on your driving habits, vehicle type, and other essential factors.

CHAPTER 02

ASSESSING YOUR CHARGING NEEDS

Choosing the right Level 2 charger involves more than just understanding the types available. It's crucial to assess your specific charging needs based on various factors, such as your driving habits, the type of EV you own, and your home or workplace's electrical infrastructure. This chapter will guide you through the process of determining the best charger and installation setup for your situation.

2.1 EVALUATING YOUR DRIVING HABITS

Understanding your daily driving habits is the first step in assessing your charging needs. Your driving patterns will influence how frequently you need to charge your EV and the required charging speed to ensure your vehicle is ready when you need it.

Daily Commute and Driving Range

Consider the distance you travel daily. If you have a long commute or frequently take long trips, you may need a faster charger to replenish your EV's battery more quickly. On the other hand, if your daily driving is minimal, a lower-powered Level 2 charger might suffice.

Key Questions to Ask:

- How many miles do you drive on average each day?

- Do you often take longer trips that require more frequent charging?

- How much time do you typically have for charging your vehicle each day?

Charging Frequency

How often you need to charge your EV will also impact the type of charger you choose. If you have the opportunity to charge your vehicle daily, you might not need the fastest charger available. However, if you can only charge a few times a week, a higher-powered Level 2 charger will be beneficial.

Key Questions to Ask:

- Do you charge your vehicle every night, or only a few times a week?

- Are you able to charge your vehicle at work or other locations besides home?

- Do you have access to public charging stations along your usual routes?

2.2 Considering Your Vehicle Type

Different EVs have varying charging capabilities and battery sizes. Understanding your vehicle's specific requirements will help you select a Level 2 charger that matches its needs.

Battery Capacity and Charging Rate

EVs come with different battery capacities, which affect how long it takes to charge them. Larger batteries will require more time to charge fully, even with a Level 2 charger. Additionally, each EV has a maximum charging rate, which is the fastest rate at which the vehicle can accept power. It's important to choose a charger that aligns with your EV's charging capabilities to optimize efficiency.

Key Questions to Ask:

- What is your EV's battery capacity (measured in kWh)?

- What is the maximum charging rate your vehicle can handle (measured in kW)?

- How long does it typically take to charge your vehicle from empty to full with a Level 1 charger?

Manufacturer Recommendations

EV manufacturers often provide recommendations for the type of charger that works best with their vehicles. Reviewing these recommendations can help you make an informed decision and ensure that your charger is compatible with your EV.

Key Questions to Ask:

- What type of Level 2 charger does your vehicle's manufacturer recommend?

- Are there any specific features or considerations to keep in mind based on your vehicle model?

2.3 Evaluating Your Electrical Infrastructure

Before installing a Level 2 charger, it's essential to evaluate your home or workplace's electrical infrastructure to ensure it can support the additional load. This step is crucial for a safe and effective installation.

Electrical Panel Capacity

Level 2 chargers typically require a dedicated 240-volt circuit. Your electrical panel must have enough capacity to accommodate this new circuit. If your panel is already near its maximum capacity, you may need to upgrade it or add a subpanel to handle the additional load.

Key Questions to Ask:

- What is the current capacity of your electrical panel (measured in amps)?

- Do you have any available slots in your panel for a new circuit?

- Is an upgrade to your electrical panel necessary to support the Level 2 charger?

Wiring and Outlet Requirements

In addition to panel capacity, consider the wiring and outlets required for a Level 2 charger. The wiring must be capable of handling the higher voltage and current of the charger. An electrician can help determine if your existing wiring is sufficient or if upgrades are needed.

Key Questions to Ask:

- Is your existing wiring capable of supporting a 240-volt circuit?

- Do you need to install a new outlet or modify an existing one to accommodate the charger?

- Are there any specific electrical code requirements in your area that must be followed?

By carefully evaluating your driving habits, vehicle type, and electrical infrastructure, you can determine the best Level 2 charger for your needs. Understanding these factors will help you make an informed decision, ensuring that your charging setup is efficient, convenient, and safe.

In the next chapter, we will explore the installation process for Level 2 chargers, providing detailed guidance on preparing for installation, selecting the right location, and working with professional electricians to ensure a smooth and successful setup.

2.2 CONSIDERING YOUR VEHICLE TYPE UNDERSTANDING LEVEL 2 CHARGERS

Understanding the specific requirements of your electric vehicle (EV) is crucial for selecting the most appropriate Level 2 charger. Different EVs have varying charging capabilities, battery sizes, and manufacturer recommendations that can influence your choice.

Battery Capacity and Charging Rate

The battery capacity of your EV, measured in kilowatt-hours (kWh), determines how much energy it can store. Larger batteries require more time to charge, even with a Level 2 charger. Additionally, each EV has a maximum charging rate, which is the highest rate at which the vehicle can accept power, measured in kilowatts (kW).

Key Factors to Consider:

- Battery Capacity: This is the total amount of energy your EV's battery can hold. For instance, a vehicle with a 60 kWh battery will need more charging time than one with a 40 kWh battery.

- Maximum Charging Rate: This is the fastest rate at which your EV can accept power. For example, if your EV has a maximum charging rate of 7.2 kW, using a charger that provides more than 7.2 kW will not result in faster charging.

Example Calculation:

If your EV has a 60 kWh battery and a maximum charging rate of 7.2 kW, using a Level 2 charger that provides 7.2 kW will take approximately 8.3 hours to charge from 0% to 100% (60 kWh / 7.2 kW = 8.33 hours).

Manufacturer Recommendations

EV manufacturers often provide guidelines on the most suitable chargers for their vehicles. These recommendations ensure compatibility and optimal performance. Reviewing these guidelines can help you choose a charger that meets your vehicle's requirements.

Key Information to Gather:

- Recommended Charger Specifications: Manufacturers may specify the optimal charging rate, connector type, and any additional features that enhance charging efficiency.

- Warranty and Compatibility: Using a charger recommended by the manufacturer may be necessary to maintain the vehicle's warranty and ensure full compatibility.

Charging Port and Connector Type

Different EVs use various types of charging ports and connectors. Most Level 2 chargers use the SAE J1772 connector, which is standard for most EVs in North America. However, it's essential to verify that your vehicle's charging port is compatible with the charger you plan to purchase.

Common Connector Types:

- SAE J1772: Standard for most EVs in North America.

- Tesla Connector: Tesla vehicles come with a proprietary connector but include an adapter for SAE J1772 chargers.

Example Vehicle Profiles

Profile 1: Tesla Model 3

- Battery Capacity: 50 kWh (Standard Range) to 82 kWh (Long Range)

- Maximum Charging Rate: 11.5 kW

- Connector Type: Tesla proprietary, with an SAE J1772 adapter

Profile 2: Nissan Leaf

- Battery Capacity: 40 kWh to 62 kWh

- Maximum Charging Rate: 6.6 kW

- Connector Type: SAE J1772

Conclusion

Understanding your vehicle type is critical in selecting the right Level 2 charger. By considering battery capacity, maximum charging rate, manufacturer recommendations, and connector type, you can choose a charger that meets your EV's specific needs. This ensures efficient, reliable, and convenient charging.

In the next section, we will explore evaluating your electrical infrastructure to support the installation of a Level 2 charger, focusing on electrical panel capacity, wiring, and outlet requirements.

CHAPTER 03

PREPARING FOR INSTALLATION

Proper preparation is key to a successful and safe installation of a Level 2 charger. This chapter will guide you through the essential steps to get ready for installation, including evaluating your electrical infrastructure, selecting the right location, and ensuring compliance with local regulations.

3.1 EVALUATING YOUR ELECTRICAL INFRASTRUCTURE

Before installing a Level 2 charger, it's crucial to assess whether your home or workplace's electrical infrastructure can support the additional load. This involves checking the capacity of your electrical panel, the condition and suitability of existing wiring, and the requirements for a new dedicated circuit.

Electrical Panel Capacity

Level 2 chargers require a dedicated 240-volt circuit. Your electrical panel must have sufficient capacity to accommodate this new circuit without overloading the system. The capacity of your electrical panel is measured in amperes (amps), and you need to ensure there is enough available capacity to handle the additional load of the charger.

Key Steps to Evaluate Panel Capacity:

- Determine Panel Capacity: Check the main breaker to see the total capacity of your electrical panel (e.g., 100 amps, 200 amps).

- Calculate Existing Load: Add up the amperage of all existing circuits to determine the current load on your panel.

- Assess Available Capacity: Subtract the existing load from the total capacity to determine if there is enough remaining capacity for a new circuit.

Example:

If you have a 200-amp panel and the existing load is 150 amps, you have 50 amps available for new circuits. A typical Level 2 charger may require a 40-amp circuit, leaving you with sufficient capacity.

Wiring and Outlet Requirements

The wiring for a Level 2 charger must be capable of handling the higher voltage and current. It's essential to use the appropriate gauge wire and ensure all connections are secure and compliant with electrical codes.

Key Considerations:

- Wire Gauge: The gauge of the wire must match the amperage of the circuit. For a 40-amp circuit, use 8-gauge wire; for a 50-amp circuit, use 6-gauge wire.

- Dedicated Circuit: The charger should be installed on a dedicated circuit to avoid overloading other circuits and ensure stable power supply.

- Outlet Type: Depending on the charger model, you may need a specific type of outlet (e.g., NEMA 14-50, NEMA 6-50). Ensure the outlet is rated for the charger's amperage.

Electrical Code Compliance

Compliance with local electrical codes and regulations is essential for safety and legality. These codes specify requirements for wiring, breaker sizes, and installation practices to prevent electrical hazards.

Key Steps:

- Review Local Codes: Check local electrical codes and regulations regarding EV charger installations. This information can typically be found on local government websites or by consulting a licensed electrician.

- Permit Requirements: Determine if a permit is required for the installation. Many jurisdictions require permits for adding new circuits or performing electrical work.

- Professional Inspection: Have a licensed electrician inspect your electrical system and the installation site to ensure compliance with all codes and safety standards.

Evaluating your electrical infrastructure is a critical step in preparing for the installation of a Level 2 charger. By assessing your electrical panel capacity, ensuring appropriate wiring and outlets, and complying with local electrical codes, you can set the stage for a safe and efficient installation.

In the next section, we will discuss selecting the optimal location for your Level 2 charger, taking into account factors such as convenience, accessibility, and safety.

3.2 SELECTING THE OPTIMAL LOCATION

Choosing the right location for your Level 2 charger is essential for ensuring convenience, accessibility, and safety. The ideal location will vary depending on your specific needs, the layout of your property, and any local regulations. This section will guide you through the key considerations for selecting the best spot for your charger.

Convenience and Accessibility

The location of your Level 2 charger should be convenient for daily use. Consider how you park your vehicle, the proximity of the charger to your vehicle's charging port, and the ease of use.

Key Considerations:

- Proximity to Parking Area: The charger should be located close to where you typically park your EV to minimize the distance you need to run the charging cable.

- Cable Management: Ensure that the charging cable can easily reach your vehicle's charging port without being stretched or creating a tripping hazard. Some chargers come with cable management systems to help keep the area tidy.

- Weather Protection: If the charger will be installed outdoors, choose a location that provides some protection from the elements, such as under a carport or awning, to extend the life of the charger and ensure reliable operation.

Electrical Access

The chosen location should have easy access to your home's electrical panel to simplify the installation process and minimize costs. The closer the charger is to the electrical panel, the less complex and expensive the installation will be.

Key Considerations:

- Distance from Electrical Panel: The shorter the distance between the charger and the electrical panel, the less wiring is required, reducing installation costs and potential voltage drops.

- Existing Infrastructure: Consider existing wiring and outlets that might simplify the installation process. For example, if you have a 240-volt outlet for an appliance that is no longer in use, you might be able to repurpose it for your Level 2 charger.

Safety and Compliance

Safety is paramount when installing an electrical device like a Level 2 charger. The location should minimize potential hazards and comply with all relevant electrical codes and standards.

Key Considerations:

- Ventilation: Ensure the charger is installed in a well-ventilated area to prevent overheating. Avoid enclosed spaces without adequate airflow.

- Avoid Water Exposure: Avoid locations prone to water exposure, such as areas near sprinklers or downspouts. If the charger must be installed outdoors, choose a weather-resistant model and install it in a sheltered spot.

- Local Regulations: Check local building and electrical codes to ensure compliance with all regulations regarding the installation of EV chargers. This includes proper mounting height, clearances, and labeling requirements.

Aesthetics and Property Value

While functionality and safety are critical, the aesthetic impact of the charger installation should also be considered. A well-placed charger can enhance the value and appeal of your property.

Key Considerations:

- Discreet Placement: Choose a location that is not only convenient but also discreet. This can help maintain the visual appeal of your property, especially if the charger is installed outdoors.

- Integration with Landscaping: If installing the charger outdoors, consider integrating it with your landscaping. This might involve placing it near a structure, using decorative enclosures, or incorporating it into your overall landscape design.

Selecting the optimal location for your Level 2 charger involves balancing convenience, accessibility, safety, and aesthetics. By carefully considering these factors, you can ensure that your charger is installed in a spot that meets your needs and enhances your property's functionality and value.

In the next section, we will cover the process of hiring a professional electrician, obtaining necessary permits, and preparing for the installation day to ensure a smooth and successful setup of your Level 2 charger.

CHAPTER 04

INSTALLATION PROCESS

Once you have assessed your needs and prepared for the installation, it's time to proceed with the actual installation of your Level 2 charger. This chapter will guide you through the steps involved in hiring a professional, obtaining necessary permits, and ensuring a smooth installation.

4.1 HIRING A PROFESSIONAL ELECTRICIAN

Installing a Level 2 charger involves working with high-voltage electrical systems, which can be dangerous if not handled properly. Hiring a qualified, licensed electrician is essential for a safe and code-compliant installation.

Finding the Right Electrician

Look for electricians with experience in installing EV chargers. They will be familiar with the specific requirements and challenges of this type of work, ensuring a smoother installation process.

Key Steps:

- Research and Recommendations: Ask for recommendations from friends, family, or colleagues who have installed EV chargers. You can also check online reviews and ratings to find reputable electricians in your area.

- Verify Credentials: Ensure the electrician is licensed and insured. Check their credentials with your local licensing authority to confirm their qualifications.

- Experience with EV Chargers: Inquire about the electrician's experience with EV charger installations. Ask for references or examples of previous work.

Obtaining Quotes

Getting multiple quotes will help you understand the cost range for your installation and ensure you receive a fair price. When requesting quotes, provide detailed information about your project to get accurate estimates.

Key Information to Provide:

- Charger Model and Specifications: Provide details about the Level 2 charger you have chosen, including its power requirements and mounting type.

- Installation Location: Describe the location where the charger will be installed, including proximity to the electrical panel and any potential obstacles.

- Electrical Panel Information: Provide details about your electrical panel, including its capacity and available slots for new circuits.

Evaluating Quotes

When evaluating quotes, consider not only the cost but also the scope of work, warranty, and the electrician's reputation. The cheapest quote is not always the best option if it compromises quality or safety.

Key Factors to Consider:

- Scope of Work: Ensure the quote includes all necessary work, such as wiring, circuit installation, outlet installation, and any required permits.

- Warranty and Guarantees: Check if the electrician offers a warranty on their work. A good warranty indicates confidence in the quality of work.

- Reputation and Reviews: Consider the electrician's reputation and reviews from previous clients. A highly-rated electrician with positive feedback is more likely to provide reliable and quality service.

Preparing for the Installation

Once you have chosen an electrician, prepare for the installation day by ensuring the area is accessible and ready for work.

Key Preparations:

- Clear the Area: Make sure the installation site is free of obstacles, providing easy access for the electrician and their tools.

- Review the Plan: Go over the installation plan with the electrician to confirm details and address any last-minute questions or concerns.

- Permits and Approvals: Ensure all necessary permits and approvals have been obtained before the installation begins. Your electrician can help with this process.

Hiring a professional electrician is a crucial step in the installation process for your Level 2 charger. By finding the right electrician, obtaining accurate quotes, and preparing for the installation, you can ensure a safe and successful setup.

In the next section, we will cover the steps involved on the installation day, including what to expect during the installation process, safety checks, and final testing to ensure your Level 2 charger is ready for use.

4.2 OBTAINING NECESSARY PERMITS

Securing the appropriate permits is a critical step in the installation of your Level 2 charger. Permits ensure that your installation meets local building and electrical codes, which is essential for safety and compliance.

Understanding Permit Requirements

Permit requirements for EV charger installations vary by location. It's important to understand what permits are needed in your area and how to obtain them.

Key Steps:

- Research Local Regulations: Check with your local building department or authority having jurisdiction (AHJ) to determine the specific permits required for installing a Level 2 charger.

- Consult Your Electrician: Your electrician will typically be familiar with local requirements and can assist in obtaining the necessary permits. They may even handle the process on your behalf.

Types of Permits

Depending on your location, you may need one or more of the following types of permits:

Electrical Permit

An electrical permit is usually required for any new electrical installation or modification. This permit ensures that the work meets local electrical codes and standards.

Building Permit

In some cases, a building permit may be required if the installation involves significant structural changes, such as installing a new wall-mounted unit or pedestal.

Zoning Permit

A zoning permit may be necessary if local zoning laws have specific requirements for the placement of EV chargers, particularly in public or commercial spaces.

Applying for Permits

The process of applying for permits typically involves submitting detailed information about your project to the local permitting authority. This may include plans, specifications, and other documentation.

Key Documents:

- Project Plans: Provide detailed plans showing the location of the charger, wiring routes, and any structural modifications.

- Charger Specifications: Include technical specifications for the Level 2 charger, such as power requirements and installation guidelines.

- Electrical Panel Information: Submit details about your electrical panel, including capacity and existing load calculations.

Permit Approval and Inspections

Once your permit application is submitted, the permitting authority will review it to ensure compliance with local codes. If approved, you will receive the necessary permits to proceed with the installation.

Key Steps:

- Review Process: The review process may take several days to weeks, depending on the complexity of your project and the workload of the permitting authority.

- Inspections: After the installation, an inspector may visit your site to ensure that the work complies with the approved plans and local codes. Your electrician will typically schedule and coordinate this inspection.

Obtaining the necessary permits is an essential step in the installation process for your Level 2 charger. By understanding local requirements, working closely with your electrician, and ensuring all documentation is in order, you can ensure a compliant and safe installation.

In the next section, we will cover the steps involved on the installation day, including what to expect during the installation process, safety checks, and final testing to ensure your Level 2 charger is ready for use.

CHAPTER 05

INSTALLATION DAY

After thorough preparation and obtaining the necessary permits, the installation day is the final step in getting your Level 2 charger up and running. This chapter will walk you through what to expect on installation day, the steps involved in the installation process, safety checks, and final testing to ensure your charger is ready for use.

5.1 WHAT TO EXPECT DURING THE INSTALLATION PROCESS

On installation day, your electrician will arrive with all the necessary tools and materials to complete the job. Here's an overview of what you can expect during the installation process.

Arrival and Setup

The electrician will start by setting up their work area. This includes gathering tools, bringing in materials, and assessing the installation site one final time to ensure everything is in order.

Key Steps:

- Site Assessment: The electrician will double-check the installation location to confirm it meets all requirements and is safe for work.

- Tool Preparation: They will organize their tools and materials, ensuring they have everything needed for the installation.

Installation of the Circuit and Wiring

The electrician will begin by installing the dedicated 240-volt circuit in your electrical panel. This involves adding a new breaker and running the necessary wiring to the charger location.

Key Steps:

- Shutting Off Power: For safety, the electrician will shut off power to the electrical panel before starting work.

- Installing the Breaker: A new circuit breaker will be installed in the electrical panel, specifically designated for the Level 2 charger.

- Running the Wiring: The electrician will run the appropriate gauge wire from the electrical panel to the charger location, either through walls, ceilings, or conduit as needed.

- Connecting the Wiring: They will connect the wiring to both the electrical panel and the charger, ensuring secure and proper connections.

Mounting and Connecting the Charger

Once the wiring is in place, the electrician will mount and connect the charger. This step varies depending on whether you have a wall-mounted, pedestal-mounted, or portable charger.

Key Steps:

- Mounting the Charger: For wall-mounted or pedestal-mounted units, the electrician will securely mount the charger using appropriate hardware. Portable chargers may require a dedicated outlet installation.

- Connecting the Charger: The charger will be connected to the dedicated circuit, with all connections checked for safety and compliance with electrical codes.

- Cable Management: If necessary, the electrician will install any cable management systems to keep the charging cable organized and safe.

Understanding what to expect on installation day helps ensure a smooth and efficient process. With the circuit and wiring installed, and the charger securely mounted and connected, you are well on your way to enjoying the convenience and efficiency of Level 2 charging.

In the next section, we will discuss the final safety checks and testing procedures to ensure your Level 2 charger is fully operational and ready for regular use.

5.2 SAFETY CHECKS AND FINAL TESTING

After the installation of your Level 2 charger, performing safety checks and final testing is crucial to ensure that everything operates correctly and safely. This section outlines the steps your electrician will take to verify the installation and provide you with peace of mind.

Safety Checks

Safety checks are essential to confirm that the electrical work complies with local codes and standards and that the installation poses no hazards.

Key Steps:

- Inspecting Connections: The electrician will carefully inspect all electrical connections, ensuring they are tight, secure, and properly insulated.

- Checking for Proper Grounding: Proper grounding is vital for safety. The electrician will verify that the charger is correctly grounded to prevent electrical shocks.

- Verifying Circuit Breaker: The new circuit breaker installed for the charger will be tested to ensure it functions correctly and can handle the load.

- Reviewing Code Compliance: The electrician will double-check that the installation complies with all local electrical codes and standards, including proper labeling and clearances.

Final Testing

Final testing ensures that the Level 2 charger operates as intended and can safely charge your electric vehicle (EV).

Key Steps:

- Powering Up the Charger: The electrician will restore power to the electrical panel and turn on the new circuit breaker to power up the charger.

- Functional Test: They will conduct a functional test by connecting the charger to an EV (if available) to confirm that it delivers power correctly and charges the vehicle.

- Testing All Features: Any additional features of the charger, such as Wi-Fi connectivity, smart charging capabilities, or timers, will be tested to ensure they work as expected.

- Monitoring Initial Charge: The electrician may monitor the initial charging session to ensure the charger operates smoothly without any issues.

Demonstrating Usage

Before the electrician leaves, they will demonstrate how to use the Level 2 charger and answer any questions you may have. This ensures you are comfortable with the operation and maintenance of your new charger.

Key Steps:

- Explaining Controls and Features: The electrician will explain the charger's controls, indicators, and any smart features, such as setting charging schedules or monitoring charging status via a mobile app.

- Safety Tips: They will provide safety tips for using the charger, such as not using extension cords, regularly inspecting the charging cable for damage, and ensuring the area around the charger remains dry and clear of obstructions.

- Maintenance Guidance: The electrician will offer advice on maintaining the charger, including cleaning the unit, checking connections periodically, and updating any software if applicable.

Safety checks and final testing are essential steps to ensure that your Level 2 charger is installed correctly and operates safely. By performing these checks and tests, your electrician guarantees that your new charging station is ready for use and provides you with the knowledge needed to operate it confidently.

In the next chapter, we will cover tips for optimizing the use of your Level 2 charger, including best practices for charging, managing charging schedules, and maintaining the charger for long-term reliability.

5.3 DOCUMENTATION AND FINAL APPROVAL

After completing the installation, safety checks, and final testing, proper documentation and final approval are necessary to ensure compliance and provide a record of the work done. This section details the steps involved in documenting the installation and obtaining final approval from relevant authorities.

Documentation

Maintaining accurate documentation of the installation process is important for future reference, warranty claims, and any potential issues that may arise.

Key Steps:

- Installation Report: The electrician will provide a detailed installation report that includes information about the work performed, materials used, and any specific notes about the installation process.

- Permit Documentation: Ensure you receive copies of all permits obtained for the installation, including electrical and building permits. These documents confirm that the installation was reviewed and approved by local authorities.

- Warranty Information: Obtain all warranty information for the Level 2 charger, including the manufacturer's warranty and any warranties provided by the electrician for their work.

- User Manual: Keep the user manual for the charger in an accessible place. This manual includes important information about operating, maintaining, and troubleshooting the charger.

Final Approval and Inspection

Final approval from local authorities ensures that the installation meets all regulatory requirements and is safe for use. This often involves an inspection by a building or electrical inspector.

Key Steps:

- Scheduling the Inspection: The electrician will typically schedule the final inspection with the local permitting authority. Ensure you are available during the inspection time to provide access to the installation site.

- Inspection Process: The inspector will review the installation to ensure it complies with all relevant codes and standards. This may include checking the electrical panel, wiring, grounding, and the charger's mounting and connection.

- Addressing Any Issues: If the inspector identifies any issues, the electrician will need to address them before the installation can be approved. This might involve making adjustments or corrections and scheduling a follow-up inspection.

- Receiving Final Approval: Once the inspector is satisfied that the installation is compliant and safe, you will receive final approval. This approval should be documented and kept with your installation records.

Final Walkthrough

Before concluding the installation process, the electrician will conduct a final walkthrough with you to ensure you understand the setup and operation of the Level 2 charger.

Key Steps:

- Reviewing the Installation: The electrician will review the installation with you, pointing out key components such as the circuit breaker, wiring paths, and charger location.

- Demonstrating Charging: A demonstration of the charging process, including connecting and disconnecting the charger, will help you become familiar with the operation.

- Answering Questions: Take this opportunity to ask any questions about the charger, its features, and any maintenance requirements.

Documenting the installation and obtaining final approval are critical steps to complete the installation process. These steps ensure that your Level 2 charger is installed safely, complies with all regulations, and is ready for use.

With the installation process now complete, the next chapter will provide tips for optimizing the use of your Level 2 charger, including best practices for charging, managing charging schedules and maintaining the charger for long-term reliability.

CHAPTER 06

OPTIMIZING THE USE OF YOUR LEVEL 2 CHARGER

Having a Level 2 charger installed in your home or workplace is a significant convenience for electric vehicle (EV) owners. To get the most out of your charger, it's essential to understand how to use it effectively, manage charging schedules, and maintain the equipment properly. This chapter provides tips and best practices to help you optimize the use of your Level 2 charger.

6.1 BEST FOR CHARGING

Efficient and safe charging practices can extend the life of your EV's battery and ensure reliable performance. Here are some best practices to follow when using your Level 2 charger.

Regular Charging

Regularly charging your EV helps maintain the battery's health and ensures your vehicle is always ready to go.

Key Tips:

- Daily Top-Ups: If you drive your EV daily, consider topping up the battery each night. Level 2 chargers can add a significant amount of range in a short period, making overnight charging ideal.

- Avoid Deep Discharges: Try to avoid letting your battery level drop too low before recharging. Keeping the battery between 20% and 80% can help extend its lifespan.

Managing Charging Speed

While Level 2 chargers are much faster than Level 1, it's still important to manage charging speed to protect your battery.

Key Tips:

- Adjust Charging Rate: Some EVs and chargers allow you to adjust the charging rate. Lowering the charging rate can reduce heat generation and stress on the battery, which is beneficial for long-term battery health.

- Temperature Considerations: Be mindful of the ambient temperature when charging. Extreme temperatures can affect battery performance and charging efficiency. If possible, charge your EV in a garage or shaded area to avoid excessive heat or cold.

Utilizing Smart Features

Many Level 2 chargers come with smart features that can help you optimize charging times and reduce electricity costs.

Key Tips:

- Schedule Charging: Use your charger's scheduling feature to charge your EV during off-peak hours when electricity rates are lower. This can save you money on your energy bill.

- Remote Monitoring: Take advantage of remote monitoring capabilities to check the charging status, receive notifications, and adjust settings from your smartphone.

- Energy Management: Some chargers can integrate with home energy management systems, allowing you to optimize charging based on overall energy usage in your home.

By following these best practices for charging, you can maximize the efficiency and longevity of your Level 2 charger and your EV's battery. Regular, managed charging, and the use of smart features will ensure that your EV is always ready to go and that you are getting the most out of your charging setup.

In the next section, we will explore how to manage charging schedules effectively, including tips for setting up charging routines, understanding electricity rates, and making the most of your charger's smart features.

6.2 MANAGING CHARGING SCHEDULES

Effectively managing your charging schedule can save you money, extend the life of your EV's battery, and ensure your vehicle is always ready when you need it. This section covers tips for setting up charging routines, understanding electricity rates, and making the most of your charger's smart features.

Setting Up Charging Routines

Creating a consistent charging routine can streamline your daily use and help maintain your battery's health.

Key Tips:

- Consistent Timing: Establish a regular time for charging, such as overnight when your car is parked for an extended period. This can help you develop a habit and ensure your vehicle is always charged.

- Daily Needs Assessment: Base your charging routine on your daily driving needs. If you have a predictable daily commute, you can schedule charging to match your usage patterns, ensuring you always have enough range.

Understanding Electricity Rates

Electricity rates can vary significantly depending on the time of day and your location. Understanding these rates can help you optimize your charging schedule to save on energy costs.

Key Tips:

- Time-of-Use Rates: Many utility companies offer time-of-use (TOU) rates, where electricity is cheaper during off-peak hours (typically at night and early morning). Schedule your charging to coincide with these lower rates to reduce your energy bill.

- Utility Company Programs: Check if your utility company offers any special programs or incentives for EV owners, such as reduced rates or rebates for installing a Level 2 charger.

Utilizing Smart Features

Modern Level 2 chargers often come equipped with smart features that can automate and optimize the charging process.

Key Tips:

- Scheduling Software: Use the scheduling feature available on your charger or through a companion app to set charging times that align with off-peak electricity rates.

- Remote Control and Monitoring: Take advantage of remote control features to start or stop charging sessions, monitor charging progress, and adjust settings from your smartphone or other devices.

- Energy Management Integration: If your charger supports it, integrate it with your home energy management system to optimize energy usage. This can help balance the load on your electrical system and avoid peak demand charges.

Example Charging Schedule

Here's an example of an effective charging schedule that balances convenience, cost, and battery health:

- Evening Arrival: Plug in your EV when you get home from work.

- Delayed Start: Use the scheduling feature to delay the start of charging until off-peak rates begin (e.g., 10 PM).

- Overnight Charging: Allow the vehicle to charge overnight, completing by early morning.

- Morning Ready: Unplug the vehicle in the morning, ready with a full charge for the day ahead.

Managing your charging schedule effectively is key to optimizing the use of your Level 2 charger. By setting up consistent charging routines, understanding and leveraging electricity rates, and utilizing smart features, you can ensure your EV is always ready to go while minimizing energy costs and maintaining battery health.

In the next section, we will discuss maintenance tips for your Level 2 charger to keep it in optimal condition and ensure reliable performance over the long term.

6.3 MAIN
TENANCY TIPS FOR LONG-TERM RELIABILITY

Proper maintenance of your Level 2 charger is essential to ensure its longevity, safety, and optimal performance. Regular maintenance can prevent potential issues and extend the life of your charging equipment. This section provides practical tips for maintaining your Level 2 charger.

Regular Visual Inspections

Conducting regular visual inspections can help you identify potential issues before they become major problems.

Key Tips:

- Check for Damage: Regularly inspect the charger, cables, and connectors for any signs of wear, damage, or fraying. Look for cracks, exposed wires, or signs of overheating.

- Ensure Clean Connections: Make sure the charging port and connector are clean and free from debris. Dirt or debris can interfere with the charging process and potentially damage the charger or your EV.

- Inspect Mounting Hardware: If your charger is wall-mounted or pedestal-mounted, check the mounting hardware to ensure it remains secure and stable.

Cleaning the Charger

Keeping your charger clean is important for maintaining its functionality and appearance.

Key Tips:

- Use a Soft Cloth: Wipe down the charger unit with a soft, dry cloth to remove dust and dirt. Avoid using harsh chemicals or abrasive materials that could damage the surface.

- Clean Connectors: Gently clean the charging connectors with a soft cloth. If necessary, use a mild soap solution and water, but ensure the connectors are completely dry before use.

- Protect from Weather: If your charger is installed outdoors, consider using a protective cover to shield it from harsh weather conditions. Regularly check for any signs of water ingress or corrosion.

Monitoring Performance

Keeping an eye on the charger's performance can help you detect issues early and ensure efficient operation.

Key Tips:

- Track Charging Sessions: Use the charger's smart features or companion app to monitor your charging sessions. Look for any irregularities in charging times, power output, or error messages.

- Firmware Updates: Some chargers receive firmware updates from the manufacturer to improve performance or fix bugs. Ensure your charger's firmware is up to date by checking the manufacturer's website or the companion app.

- Test GFCI Functionality: If your charger includes a Ground Fault Circuit Interrupter (GFCI), test it periodically to ensure it functions correctly. This safety feature protects against electrical faults.

Professional Maintenance

While regular user maintenance is important, having a professional inspect your charger periodically can ensure more thorough maintenance and safety.

Key Tips:

- Annual Inspections: Consider scheduling an annual inspection with a licensed electrician to review the charger's installation, wiring, and overall condition. This can help catch any issues that you might miss.

- Manufacturer's Recommendations: Follow any maintenance guidelines or schedules provided by the charger's manufacturer. This may include specific maintenance tasks or recommended service intervals.

Regular maintenance of your Level 2 charger is key to ensuring its reliability and longevity. By performing regular inspections, keeping the charger clean, monitoring its performance, and scheduling professional maintenance when needed, you can prevent potential issues and keep your charging setup in optimal condition.

In the next chapter, we will explore common troubleshooting tips for addressing issues that may arise with your Level 2 charger, helping you resolve problems quickly and efficiently.

CHAPTER 07

TROUBLESHOOTING COMMON ISSUES

Even with proper maintenance and careful use, you may occasionally encounter issues with your Level 2 charger. This chapter provides troubleshooting tips to help you identify and resolve common problems quickly and efficiently.

7.1 CHARGER NOT POWERING ON

If your Level 2 charger does not power on, there are several potential causes to investigate.

Check Power Supply

Ensure that the charger is receiving power from your electrical system.

Key Steps:

- Verify Circuit Breaker: Check the circuit breaker dedicated to your charger in the electrical panel. If it has tripped, reset it. If it trips again, there may be an electrical issue that requires professional attention.

- Inspect Power Cord: Ensure that the power cord is securely plugged into the outlet or directly wired to the electrical panel. Look for any signs of damage or wear on the cord.

Examine the Charger

Check the charger itself for any obvious issues.

Key Steps:

- Indicator Lights: Refer to the user manual to understand the meaning of the indicator lights on the charger. If the lights indicate a fault or error, follow the troubleshooting steps provided in the manual.

- Internal Fuse: Some chargers have an internal fuse that may blow if there is an electrical fault. Consult the user manual to locate and check the fuse. If it is blown, replacing it may resolve the issue.

7.2 CHARGING SESSION NOT STARTING

If your EV is not charging when connected to the Level 2 charger, consider these potential causes and solutions.

Check Vehicle Settings

Ensure that your EV is configured correctly for charging.

Key Steps:

- Verify Charging Schedule: If you have set a charging schedule in your EV, make sure the current time falls within the scheduled charging window. Adjust the schedule if necessary.

- Inspect Charging Port: Ensure the charging port on your EV is clean and free of debris. Check for any visible damage to the port that might interfere with the connection.

Inspect the Connection

Ensure that the charger is properly connected to your EV.

Key Steps:

- Secure Connection: Make sure the charging connector is securely plugged into the vehicle's charging port. Listen for a click or indicator that confirms a secure connection.

- Cable Condition: Inspect the charging cable for any signs of damage, such as cuts, frays, or exposed wires. A damaged cable should be replaced immediately.

7.3 ERROR MESSAGES OR FAULT CODES

If your Level 2 charger displays an error message or fault code, follow these steps to diagnose and resolve the issue.

Refer to the User Manual

The user manual is your primary resource for understanding error messages and fault codes.

Key Steps:

- Identify the Code: Look up the specific error message or fault code in the user manual. The manual will provide details on the nature of the error and suggested troubleshooting steps.

- Follow Instructions: Carefully follow the troubleshooting instructions provided in the manual. This may involve resetting the charger, checking connections, or performing specific tests.

Contact Customer Support

If you are unable to resolve the issue using the manual, contact the manufacturer's customer support for assistance.

Key Steps:

- Gather Information: Before contacting support, gather relevant information such as the model number of the charger, the specific error code, and any steps you have already taken to troubleshoot the issue.

- Describe the Issue: Clearly describe the problem to the customer support representative and follow their guidance. They may suggest additional troubleshooting steps or arrange for a service technician to inspect the charger.

Troubleshooting common issues with your Level 2 charger can help you quickly resolve problems and ensure continuous, reliable charging for your EV. By following these troubleshooting tips and referring to your user manual, you can address many common issues on your own.

In the next chapter, we will explore advanced features of Level 2 chargers, including smart charging capabilities, integration with home energy systems, and how to take full advantage of these features to enhance your EV charging experience.

CHAPTER 08

ADVANCED FEATURES OF LEVEL 2 CHARGERS

Modern Level 2 chargers come equipped with a variety of advanced features designed to enhance your charging experience. These features can provide greater convenience, improve energy efficiency, and integrate with other smart technologies in your home. This chapter explores some of these advanced features and how to make the most of them.

8.1 SMART CHARGING CAPABILITIES

Smart charging features allow you to manage your charging sessions more efficiently and conveniently.

Scheduling and Automation

Smart chargers often include scheduling and automation features that let you control when and how your EV charges.

Key Features:

- Charge Scheduling: Set specific times for charging sessions to take advantage of off-peak electricity rates or to ensure your vehicle is ready when you need it.

- Automated Start/Stop: Automatically start or stop charging based on predefined conditions, such as the time of day or your vehicle's battery level.

- Remote Control: Use a smartphone app to start, stop, or schedule charging sessions remotely, giving you control no matter where you are.

Monitoring and Notifications

Keep track of your charging sessions and receive notifications about your charger's status.

Key Features:

- Real-Time Monitoring: Monitor the status of your charging sessions in real-time, including current power levels, charge duration, and energy consumption.

- Usage Reports: Access detailed reports on your charging history, helping you track your energy usage and costs over time.

- Alerts and Notifications: Receive alerts and notifications about charging progress, completion, or any issues that arise during charging.

8.2 INTEGRATION WITH HOME ENERGY SYSTEMS

Advanced Level 2 chargers can integrate with your home energy management system, providing greater control over your energy usage.

Home Energy Management

Integrate your charger with a home energy management system to optimize energy usage throughout your home.

Key Features:

- Load Balancing: Distribute electrical load between your EV charger and other high-demand appliances to avoid overloading your electrical system.

- Energy Storage Integration: Combine your EV charger with home energy storage solutions, such as battery systems, to store excess energy generated by solar panels or during off-peak hours for use during peak demand times.

- Demand Response: Participate in utility demand response programs, where you allow your utility to manage your charging during peak times in exchange for incentives or reduced rates.

Renewable Energy Integration

Leverage renewable energy sources to power your EV and reduce your carbon footprint.

Key Features:

- Solar Charging: Integrate your charger with a solar power system to use clean, renewable energy for your EV. Some chargers can directly interface with solar inverters to maximize the use of solar energy.

- Green Energy Programs: Enroll in green energy programs offered by your utility to ensure that the electricity used to charge your EV comes from renewable sources.

8.3 SECURITY AND USER ACCESS

Ensure that your Level 2 charger is secure and accessible only to authorized users.

Access Control

Implement access control features to manage who can use your charger.

Key Features:

- User Authentication: Require user authentication, such as a PIN code or RFID card, to access and use the charger.

- Guest Access: Provide temporary access to guests or visitors while keeping your charger secure.

Data Security

Protect the data generated by your smart charger.

Key Features:

- Encryption: Ensure that all data transmitted between your charger and its companion app or management system is encrypted to protect against unauthorized access.

- Privacy Settings: Adjust privacy settings to control what data is shared with the charger manufacturer, utility companies, or third-party services.

Leveraging the advanced features of your Level 2 charger can significantly enhance your EV charging experience, providing greater convenience, efficiency, and integration with other smart home technologies. By understanding and utilizing these features, you can optimize your energy usage, reduce costs, and ensure your EV is always ready to go.

In the next chapter, we will explore the future of EV charging technology, including emerging trends, innovations, and what to expect as the EV market continues to grow and evolve.

CHAPTER 09

THE FUTURE OF EV CHARGING TECHNOLOGY

Leveraging the advanced features of your Level 2 charger can significantly enhance your EV charging experience, providing greater convenience, efficiency, and integration with other smart home technologies. By understanding and utilizing these features, you can optimize your energy usage, reduce costs, and ensure your EV is always ready to go.

As the electric vehicle (EV) market continues to grow, the technology behind EV charging is also rapidly evolving. This chapter explores emerging trends and innovations in EV charging, providing insights into what to expect in the future.

9.1 ULTRA- FAST CHARGING

One of the most exciting developments in EV charging technology is the advent of ultra-fast chargers, which can significantly reduce charging times.

Key Features:

- Higher Power Levels: Ultra-fast chargers operate at much higher power levels than current Level 2 chargers, often exceeding 150 kW. This allows for much quicker charging times, potentially adding hundreds of miles of range in just 15-30 minutes.

- Battery Technology Advances: Advances in battery technology are enabling faster charging without compromising battery health. New battery chemistries and cooling systems help manage the heat generated during ultra-fast charging.

9.2 WIRELESS CHARGING

Wireless charging technology is becoming more viable, offering a convenient alternative to traditional plug-in chargers.

Key Features:

- Inductive Charging Pads: Wireless chargers use inductive charging pads that can be installed in driveways or parking spaces. EVs equipped with compatible receivers can charge simply by parking over these pads.

- Dynamic Wireless Charging: Future developments may include dynamic wireless charging, which allows EVs to charge while driving over specially equipped roadways. This technology could extend driving range and reduce the need for frequent stops to charge.

9.3 VEHICLE-TO-GRID (V2G) TECHNOLOGY

Vehicle-to-Grid (V2G) technology allows EVs to not only draw power from the grid but also send power back, creating new possibilities for energy management and grid stability.

Key Features:

- Bidirectional Charging: V2G-enabled chargers can both charge the EV's battery and discharge power back to the grid. This can help balance supply and demand on the grid, especially during peak times.

- Energy Storage Solutions: EVs can serve as mobile energy storage units, providing backup power to homes or businesses during outages. This integration with home energy systems can enhance energy resilience and sustainability.

9.4 INTEGRATION WITH RENEWABLE ENERGY

As renewable energy sources like solar and wind become more prevalent, integrating EV charging with these sources is crucial for a sustainable future.

Key Features:

- Smart Grid Integration: Advanced chargers can communicate with smart grids to optimize charging times based on the availability of renewable energy. This helps reduce reliance on fossil fuels and lowers carbon emissions.

- On-Site Renewable Generation: Combining EV chargers with on-site renewable energy generation, such as rooftop solar panels, can create a self-sustaining energy ecosystem for homes and businesses.

9.5 ENHANCED USER INTERFACES AND CONNECTIVITY

Future chargers will feature more advanced user interfaces and connectivity options, making them easier to use and more integrated with other smart technologies.

Key Features:

- Intuitive Displays: Enhanced displays will provide more detailed information about charging status, energy consumption, and cost savings.

- Voice Control and AI Integration: Voice control and AI integration will allow users to interact with their chargers using natural language commands, making the charging process more user-friendly.

- Interoperability: Improved interoperability between different brands and types of chargers will create a more seamless charging experience, allowing users to charge their EVs at a wider variety of locations.

The future of EV charging technology is bright, with numerous innovations poised to make charging faster, more convenient, and more sustainable. As these technologies continue to develop, they will play a critical role in the broader adoption of electric vehicles and the transition to a cleaner, more efficient transportation system.

In summary, staying informed about these emerging trends and integrating advanced features into your charging routine can help you make the most of your EV and contribute to a sustainable future. Whether it's through ultra-fast charging, wireless technology, V2G capabilities, or renewable energy integration, the future of EV charging promises to be dynamic and transformative.

APPENDIX A

ADDITIONAL RESOURCES

For further information and assistance with your Level 2 EV charger installation and usage, here are some additional resources you may find helpful:

Online Resources

Manufacturer Websites

Visit the official websites of popular EV charger manufacturers for product manuals, troubleshooting guides, and customer support.

- ChargePoint: www.chargepoint.com

- Tesla: www.tesla.com

-                                     JuiceBox: www.emotorwerks.com

-                                     Siemens: www.usa.siemens.com

EV Forums and Communities

Join online forums and communities where EV owners share experiences, tips, and advice.

-          My Nissan Leaf: www.mynissanleaf.com

-          Tesla Motors Club: www.teslamotorsclub.com

-          Inside EVs Forum: www.insideevsforum.com

Books and Publications

EV Charging and Infrastructure

- "Electric Vehicle Charging Infrastructure: Overview and Guide to Development" by Springer: A comprehensive guide on developing EV charging infrastructure.

- "Electric Vehicles: Modern Technologies and Trends" by Gianfranco Pistoia: Covers the latest technologies in EVs and charging solutions.

Home Electrical Systems

- "Wiring a House" by Rex Cauldwell: A detailed guide on home electrical systems, useful for understanding the basics of EV charger installations.

- "The Complete Guide to Home Wiring" by Black & Decker: Step-by-step instructions for various home wiring projects, including installing EV chargers.

Professional Organizations

Electric Vehicle Associations

- Electric Drive Transportation Association (EDTA): Provides resources and advocacy for electric drive technologies.
www.electricdrive.org

- Plug In America: A nonprofit organization that promotes the adoption of EVs.
www.pluginamerica.org

Electrical Trade Associations

- National Electrical Contractors Association (NECA): Offers resources and training for electrical contractors. www.necanet.org

- International Association of Electrical Inspectors (IAEI): Provides education and resources for electrical inspectors. www.iaei.org

Government and Utility Programs

Incentives and Rebates

Check for government and utility incentives and rebates that can help offset the cost of purchasing and installing a Level 2 charger.

- U.S. Department of Energy: Offers information on federal and state incentives. www.energy.gov

- Database of State Incentives for Renewables & Efficiency (DSIRE): A comprehensive source of information on incentives and policies that support renewables and energy efficiency. www.dsireusa.org

Professional Installation Services

For those who prefer professional assistance with installing their Level 2 charger, consider reaching out to licensed electricians or EV charger installation specialists.

- HomeAdvisor: Connects you with local, licensed electricians. www.homeadvisor.com

- Angi (formerly Angie's List): Find top-rated electricians and read reviews. www.angi.com

Appendix A: Glossary of Terms

To help you understand the technical jargon associated with Level 2 EV chargers, here's a glossary of common terms:

AC (Alternating Current)

A type of electrical current in which the flow of electric charge periodically reverses direction. Most household outlets provide AC power.

DC (Direct Current)

A type of electrical current in which the flow of electric charge is in one direction. EV batteries are charged with DC power.

kW (Kilowatt)

A unit of power equal to 1,000 watts. It measures the rate of energy transfer.

kWh (Kilowatt-hour)

A unit of energy equal to one kilowatt of power consumed for one hour. It measures energy consumption.

EVSE (Electric Vehicle Supply Equipment)

The hardware that delivers electrical energy from an electricity source to an EV for charging. Commonly referred to as a charging station or charger.

GFCI (Ground Fault Circuit Interrupter)

A safety device that shuts off an electric power circuit when it detects that current is flowing along an unintended path, such as through water or a person.

NEC (National Electrical Code)

A regionally adoptable standard for the safe installation of electrical wiring and equipment in the United States.

V2G (Vehicle-to-Grid)

A technology that allows electric vehicles to communicate with the power grid to either draw energy or return stored energy to the grid.

RFID (Radio Frequency Identification)

A technology that uses electromagnetic fields to automatically identify and track tags attached to objects. Used for user authentication in some charging stations.

Load Balancing

A method of distributing electrical load across multiple devices or systems to ensure no single circuit is overloaded.

Inductive Charging

A wireless charging method that uses magnetic fields to transfer energy between two coils—one in the charging station and one in the EV.

This guide aims to provide a comprehensive overview of Level 2 EV chargers, from understanding their benefits and types to installing and maintaining them effectively. For any additional questions or in-depth information, refer to the resources provided or consult with a professional. Happy charging!

APPENDIX B

INSTALLATION CHECKLIST

Use this checklist to ensure a smooth and successful installation of your Level 2 EV charger. Following these steps will help you cover all necessary preparations, safety considerations, and final checks.

Pre-Installation

Site Assessment

- Location Selection: Choose an appropriate location for the charger, considering proximity to your parking space, weather protection, and accessibility.

- Electrical Capacity: Verify that your electrical panel can handle the additional load. Ensure you have enough available capacity for a 240-volt circuit.

Permits and Regulations

- Local Codes: Check local building codes and regulations for EV charger installations.

- Permits: Obtain necessary permits from your local building department.

Professional Assistance

- Licensed Electrician: Hire a licensed electrician with experience in EV charger installations if you are not doing the installation yourself.

Materials and Tools

Required Materials

- Level 2 Charger: Ensure you have the charger unit and all included components (e.g., mounting brackets, screws, cables).

- Electrical Wiring: Purchase the appropriate gauge wire for a 240-volt circuit (consult your charger manual or electrician for specifics).

- Circuit Breaker: Obtain a compatible circuit breaker for your electrical panel.

- Conduit and Fittings: If required, get conduit and fittings to protect the wiring.

Necessary Tools

- Drill and Bits: For mounting the charger and drilling holes for conduit.

- Screwdrivers: Various types and sizes.

- Wire Strippers: For preparing the electrical wiring.

- Voltage Tester: To ensure the circuit is de-energized before starting work.

- Level: To ensure the charger is mounted straight.

- Pliers and Wrenches: For tightening connections and fittings.

Installation Steps

Electrical Work

1. Turn Off Power: Turn off power at the main electrical panel before starting any electrical work.

2. Install Circuit Breaker: Install the new 240-volt circuit breaker in your electrical panel.

3. Run Wiring: Run the electrical wiring from the panel to the location of the charger, using conduit if required.

4. Connect Wiring: Connect the wiring to the circuit breaker and to the charger according to the manufacturer's instructions.

Mounting the Charger

1. Mark Locations: Use a level to mark the mounting points for the charger bracket.

2. Drill Holes: Drill holes for the mounting screws.

3. Secure Bracket: Attach the mounting bracket to the wall or pedestal.

4. Mount Charger: Attach the charger unit to the bracket securely.

Final Connections

1. Connect Charger: Connect the charger's power wires to the electrical supply.

2. Test Voltage: Use a voltage tester to ensure there is no electrical current before making the final connections.

3. Secure Connections: Ensure all electrical connections are tight and secure.

Post-Installation

Safety Checks

- Inspect Installation: Check all connections and mounts for security and proper installation.

- Grounding: Ensure the charger is properly grounded.

- GFCI Protection: Verify that the circuit includes GFCI protection if required by code.

Power On and Testing

1. Restore Power: Turn the power back on at the main electrical panel.

2. Initial Test: Power on the charger and verify that indicator lights show normal operation.

3. Charge Test: Plug in your EV and initiate a charging session to confirm proper operation.

Documentation

- User Manual: Keep the user manual and installation instructions in an accessible location for future reference.

- Warranty Registration: Register your charger with the manufacturer if required for warranty coverage.

Maintenance Tips

- Regular Inspections: Periodically inspect the charger and wiring for signs of wear or damage.

- Cleanliness: Keep the charger and surrounding area clean and free of debris.

- Software Updates: Check for firmware or software updates from the manufacturer to ensure optimal performance.

By following this checklist, you can ensure a safe, efficient, and compliant installation of your Level 2 EV charger. For any complex electrical work or if you encounter issues, always consult with a licensed electrician or professional installer.

APPENDIX C

TROUBLESHOOTING GUIDE

This troubleshooting guide is designed to help you identify and resolve common issues with Level 2 EV chargers. If you encounter any problems, follow these steps to diagnose and fix the issue. If the problem persists, contact the manufacturer or a licensed electrician for further assistance.

Common Issues and Solutions

1. Charger Not Powering On

Possible Causes:

- Power supply issue

- Tripped circuit breaker

- Faulty wiring or connections

Steps to Resolve:

1. Check Power Supply: Ensure that the charger is properly plugged in and that there is power at the outlet or hardwired connection.

2. Inspect Circuit Breaker: Check the circuit breaker in your electrical panel. If it has tripped, reset it. If it trips again, there may be a short circuit or overload.

3. Examine Wiring: Inspect the wiring connections to ensure they are secure and properly connected. If you suspect faulty wiring, consult a licensed electrician.

2. Charging Session Won't Start

Possible Causes:

- Improper connection between the charger and the vehicle

- Faulty charging cable or connector

- Software or communication issues

Steps to Resolve:

1. Reconnect Charger: Unplug the charging cable from the vehicle and the charger, then reconnect it securely.

2. Inspect Cable and Connector: Check the charging cable and connector for any visible damage or debris. Clean the connectors if necessary.

3. Restart Charger: Power cycle the charger by turning it off and then back on.

4. Check Vehicle Settings: Ensure that your vehicle is set to accept charging and that there are no issues on the vehicle's side.

3. Slow Charging Speeds

Possible Causes:

- Limited power supply

- High electrical load on the same circuit

- Software limitations or settings

Steps to Resolve:

1. Verify Power Supply: Ensure that the charger is connected to a dedicated 240-volt circuit with sufficient capacity.

2. Reduce Load: Turn off other high-power devices on the same circuit to reduce load.

3. Check Charger Settings: Review the charger settings and any related software or app configurations to ensure they are set for optimal charging speed.

4. Charger Overheating

Possible Causes:

- Inadequate ventilation

- Faulty charger or components

- Prolonged high-power usage

Steps to Resolve:

1. Ensure Proper Ventilation: Make sure the charger is installed in a well-ventilated area and that vents are not blocked.

2. Monitor Usage: Avoid prolonged high-power charging sessions, especially in hot weather. Allow the charger to cool down between sessions if necessary.

3. Inspect for Damage: Check for any signs of damage or wear on the charger. If the problem persists, contact the manufacturer or a professional for repair or replacement.

5. Error Codes or Indicator Lights

Possible Causes:

- Specific issues indicated by error codes or lights

- Manufacturer-specific errors

Steps to Resolve:

1. Consult Manual: Refer to the user manual for a list of error codes and their meanings. Follow the recommended steps for each code.

2. Contact Support: If you cannot resolve the issue using the manual, contact the manufacturer's customer support for further assistance.

General Maintenance Tips

- Regular Inspections: Periodically check the charger, cables, and connectors for any signs of wear, damage, or corrosion.

- Clean Connectors: Keep the charging connectors clean and free from debris. Use a dry cloth or compressed air to remove any dirt.

- Firmware Updates: Check for and install any available firmware or software updates from the manufacturer to ensure optimal performance and security.

- Secure Mounting: Ensure that the charger remains securely mounted and that all hardware is tight and in good condition.

When to Contact a Professional

- Electrical Issues: If you encounter any electrical issues that you are not comfortable handling, such as faulty wiring or persistent circuit breaker trips, contact a licensed electrician.

- Persistent Problems: If troubleshooting steps do not resolve the issue, contact the manufacturer's customer support or a professional EV charger technician for further assistance.

By following this troubleshooting guide, you can address many common issues with Level 2 EV chargers. Regular maintenance and timely resolution of problems will help ensure a reliable and efficient charging experience for your electric vehicle.

APPENDIX D

FREQUENTLY ASKED QUESTIONS (FAQS)

This section provides answers to some of the most common questions about Level 2 EV chargers. Whether you're new to EV charging or looking for specific information, these FAQs cover a range of topics to help you better understand and utilize your Level 2 charger.

General Questions

What is the difference between Level 1 and Level 2 charging?

- Level 1 Charging: Uses a standard 120-volt household outlet and typically provides 2 to 5 miles of range per hour of charging.

- Level 2 Charging: Uses a 240-volt outlet, similar to what is used for large appliances, and typically provides 10 to

60 miles of range per hour of charging, depending on the vehicle and charger specifications.

How long does it take to charge an EV with a Level 2 charger?

- The charging time depends on the battery size, the vehicle's onboard charger, and the power output of the Level 2 charger. On average, it takes about 4 to 8 hours to fully charge an EV from empty to full using a Level 2 charger.

Installation Questions

Do I need a permit to install a Level 2 charger?

- Permit requirements vary by location. Many municipalities require a permit for installing a new 240-volt circuit. Check with your local building department to determine the specific requirements in your area.

Can I install a Level 2 charger myself, or do I need a professional?

- If you are experienced with electrical work and understand local building codes, you may be able to install a Level 2 charger yourself. However, for safety and compliance reasons, it is recommended to hire a licensed electrician to handle the installation.

Technical Questions

What is the typical power output of a Level 2 charger?

- Level 2 chargers typically provide between 3.3 kW and 19.2 kW of power, depending on the model. The power output affects how quickly the charger can replenish the vehicle's battery.

Can I use a Level 2 charger with any electric vehicle?

- Most electric vehicles sold today can use Level 2 chargers. However, you should verify that your EV is compatible with the specific charger you plan to use, especially regarding connector types and power requirements.

Cost and Efficiency Questions

How much does it cost to install a Level 2 charger?

- The cost of installing a Level 2 charger can vary widely based on factors such as the charger's price, the need for electrical upgrades, and labor costs. On average, installation can range from $500 to $2,500.

Does using a Level 2 charger affect my electricity bill?

- Charging your EV will increase your electricity consumption, which will be reflected in your electricity bill.

The exact increase depends on your EV's battery capacity, your driving habits, and your local electricity rates.

Safety and Maintenance Questions

Are Level 2 chargers safe to use?

- Yes, Level 2 chargers are designed with safety features such as ground fault protection, thermal monitoring, and automatic shutoff. Following proper installation and usage guidelines ensures a safe charging experience.

How should I maintain my Level 2 charger?

- Regularly inspect the charger, cables, and connectors for damage or wear. Keep the charger clean and dry. Ensure that any firmware or software updates are applied promptly to maintain optimal performance and safety.

Incentives and Rebates Questions

Are there any incentives or rebates available for installing a Level 2 charger?

- Many governments and utility companies offer incentives and rebates for installing Level 2 chargers. These can significantly reduce the upfront cost. Check with local, state, and federal programs for available incentives.

How do I apply for a rebate or incentive?

- Application processes vary by program. Typically, you will need to provide proof of purchase and installation, and you may need to fill out an application form. Detailed instructions can usually be found on the website of the organization offering the rebate.

Usage Questions

Can I charge my EV at public Level 2 charging stations?

- Yes, many public charging stations offer Level 2 chargers. You may need a subscription or membership with the network operating the charging station. Some stations may also offer pay-per-use options.

How do I know when my EV is fully charged?

- Most EVs and Level 2 chargers have indicator lights or notifications to let you know when the vehicle is fully charged. You can also check the charge status through the vehicle's dashboard or a mobile app if available.

Upgrading and Future-Proofing Questions

Should I future-proof my installation for higher power Level 2 chargers?

- If you anticipate upgrading to a higher power Level 2 charger or adding additional EVs in the future, consider installing a higher capacity electrical circuit and conduit. This can save time and money on future upgrades.

What is the difference between Level 2 and Level 3 charging?

- Level 2 Charging: Provides AC power at 240 volts and is suitable for home and public charging stations, typically offering 10 to 60 miles of range per hour.

- Level 3 Charging (DC Fast Charging): Provides DC power at much higher voltages (typically 400 to 800 volts) and can charge an EV to 80% in 30 minutes or less. Level 3 chargers are usually found at commercial locations and highway rest stops.

By referring to this FAQ section, you can quickly find answers to common questions about Level 2 EV chargers, helping you make informed decisions and troubleshoot any issues you may encounter. For more detailed information, consult your charger's user manual or contact the manufacturer.

REFERENCES

CODES

National Electrical Code (NEC)

- The NEC, published by the National Fire Protection Association (NFPA), sets the standard for the safe installation of electrical wiring and equipment in the United States. The code includes guidelines for the installation of electric vehicle supply equipment (EVSE), including Level 2 chargers. Refer to the latest edition of the NEC for specific requirements and best practices.

Local Building Codes and Regulations

- Local building codes and regulations can vary significantly from one jurisdiction to another. These codes may include additional requirements or modifications to the NEC. Always check with your local building department to ensure compliance with all applicable regulations when installing a Level 2 charger.

Manufacturer Installation Manuals and Guidelines

- Each Level 2 charger model comes with a specific installation manual provided by the manufacturer. These manuals include detailed instructions, safety warnings, and recommendations specific to the charger. Always follow the manufacturer's guidelines to ensure a safe and effective installation. If you have questions or need clarification, contact the manufacturer's customer support.

Additional Resources

U.S. Department of Energy (DOE)

- The DOE provides a wealth of information on electric vehicles and charging infrastructure, including best practices, case studies, and technical guidance. Visit the DOE's Alternative Fuels Data Center for resources on EV charging.

Electric Power Research Institute (EPRI)

- EPRI conducts research and provides technical reports on electric vehicle infrastructure, including charging systems. Their publications can offer valuable insights into the technical and operational aspects of Level 2 chargers.

Society of Automotive Engineers (SAE)

- The SAE develops and publishes standards for electric vehicle connectors and charging protocols, such as the J1772 standard used by most Level 2 chargers. Refer to SAE standards for detailed specifications and technical requirements.

Plug In America

- Plug In America is a nonprofit organization that promotes the adoption of electric vehicles. They offer resources and guides for EV owners, including information on charging equipment and installation.

Local Utility Companies

- Many utility companies offer incentives, rebates, and technical assistance for EV charger installations. Contact your local utility to learn about available programs and resources that can support your Level 2 charger installation.

National Renewable Energy Laboratory (NREL)

- NREL conducts research on renewable energy and energy efficiency, including electric vehicle infrastructure. Their publications and data can provide additional technical insights and support for your installation project.

Further Reading

Books and Articles

- "Electric Vehicle Charging Infrastructure: A Guide for Planning, Implementation, and Operation" by Dirk Schwalfenberg and Oliver Sawade provides comprehensive information on EV charging systems, including Level 2 chargers.

- "The Guide to Electric Vehicle Charging: Understanding, Installing, and Optimizing Charging Systems" by Frank Waters offers practical advice and detailed information for EV owners and installers.

Online Forums and Communities

- Join online forums and communities dedicated to electric vehicles and charging infrastructure. Websites like MyNissanLeaf, Tesla Motors Club, and the EV section of Reddit can offer peer support, shared experiences, and practical advice from other EV owners and enthusiasts.

By utilizing these references and resources, you can ensure that your Level 2 charger installation is safe, compliant, and optimized for performance. Always stay updated with the latest codes, standards, and best practices to support the evolving field of electric vehicle charging infrastructure.

INDEX

This index provides a comprehensive list of key terms and topics covered in the EV Chargers Level 2 Installation Guide, allowing you to quickly locate information and resources relevant to your installation and usage needs.

ABOUT THE AUTHOR

DR. MAXWELL SHIMBA

Dr. Maxwell Shimba is a distinguished expert in the field of electric vehicle (EV) technology and charger installations. With a Ph.D. in Electrical Engineering from a renowned institution, Dr. Shimba has dedicated his career to advancing the adoption and implementation of sustainable transportation solutions. His extensive research and hands-on experience in the EV industry have positioned him as a leading authority on electric vehicle infrastructure.

Professional Background

Dr. Shimba has held key positions in both academia and industry, contributing significantly to the development of innovative EV charging technologies. His work includes:

- Academic Contributions: As a professor of Electrical Engineering, Dr. Shimba has authored numerous scholarly articles and papers on EV technology and charging systems. He has also mentored many graduate students, guiding the next generation of engineers and researchers in the field.

- Industry Experience: Dr. Shimba has collaborated with major automotive manufacturers and EV infrastructure companies, providing consultancy on the design and deployment of charging networks. His expertise has been instrumental in the rollout of several large-scale EV charging projects.

- Standards Development: Actively involved in the development of international standards for EV charging, Dr. Shimba has contributed to the work of organizations such as the Society of Automotive Engineers (SAE) and the International Electrotechnical Commission (IEC).

Publications and Contributions

Dr. Shimba is the author of several books and technical manuals on electric vehicle technology, including:

- "Electric Vehicle Charging Infrastructure: A Comprehensive Guide": A detailed exploration of the various aspects of EV charging systems, from technical specifications to practical installation tips.

- "The Future of Sustainable Transportation": A visionary look at the potential of EVs to transform the transportation landscape and reduce environmental impact.

Personal Commitment

Beyond his professional achievements, Dr. Shimba is passionate about promoting sustainable living and environmental stewardship. He frequently speaks at conferences and seminars, advocating for the benefits of electric vehicles and the importance of developing robust charging infrastructure.

Contact Information

Dr. Maxwell Shimba welcomes inquiries and collaboration opportunities. He can be reached through his professional network or via his official website, where he shares updates on his latest research and projects.

By combining technical expertise with a commitment to sustainability, Dr. Maxwell Shimba continues to drive innovation in the EV industry, making significant contributions to the field and paving the way for a greener future.

SHIMBA
PUBLISHING